Scientific Songs of Praise
Humanist Hymns and Secular Christmas Carols

Dr T. Abdulla

ISBN: 978-1517395261
ISBN-13: 1517395267

CONTENTS

1. Let's Do Science (Deck The Halls)

Gather your ideas eclectic
Fa la la la la la la la la la
'Tis the season to be sceptic
Fa la la la la la la la la la
Build theories from observation
Fa la la la la la la la la la
Be open to falsification
Fa la la la la la la la la la

Pursue your thoughts with great persistence
Fa la la la la la la la la la
Probe the questions of existence
Fa la la la la la la la la
Design a test and carefully measure
Fa la la la la la la la la
Make new findings, feel great pleasure
Fa la la la la la la la la

Reconcile the contradictions
Fa la la la la la la la la
Build models and make predictions
Fa la la la la la la la la
Standing on shoulders of giants
Fa la la la la la la la la.
Don your labcoats; Let's do Science!
Fa la la la la la la la la.

2. All Things Bright and Beautiful

All things bright and beautiful
All species big and small
All things weird and wonderful
Our world evolved them all

And very small mutations
Accumulate in time
Evolving complex forms of life
From prehistoric slime

All things bright and beautiful
All species big and small
All things weird and wonderful
Our world evolved them all

Each flower's glowing colours
Attract a bird or bee
Reflecting ultraviolet light
That humans cannot see

All things bright and beautiful
All species big and small
All things weird and wonderful
Our world evolved them all

Some plants evolved the tasty fruit
That animals love to chew
And so distribute far and wide
The seeds left in their poo

2

All things bright and beautiful
All species big and small
All things weird and wonderful
Our world evolved them all

Each sea cow and each dugong
Each whale and manatee
Evolved from ancient land based mammals
That returned to the sea

All things bright and beautiful
All species big and small
All things weird and wonderful
Our world evolved them all

The newly forming island
The river running through
Create natural barriers
And life evolves anew

All things fast and lumbering
Each predator and prey
All things quiet and thundering
They still evolve today

3. O Satellite (O Holy Night)

Oh Satellite, the stars are brightly gleaming
The moon is full as you fall round the Earth
Under the clouds, while half the world is dreaming
You're probing questions of intrigue and worth

A thrill of hope, the joy of exploration
The world is spinning an ever-breaking dawn
Orbit in peace
No fear, no segregation

In flight sublime!
Oh in flight; no longer torn.
In flight sublime!
In flight, in flight sublime.

All we can do now is love one another,
Our world's so small in this vast universe
Borders we'll take down and peace rediscover
In the aim of science, all barriers traversed

Sweet rays of sun power your onboard systems
Let all within you survive in outer space
With renewed hope
Our wavering future glistens.

No ill, no ill
In flight, in flight sublime
No ill, no ill
In flight, in flight sublime
No ill, no ill
In flight, in flight sublime.

4. A New Theory of Everything
(Hark The Herald Angels Sing)

Quarks and membranes, superstrings
A new theory of everything.
Many deep ideas compiled,
Large and small now reconciled.
Nothing can stand in our paths,
Once we understand the maths.
Universe at last revealed
In a transcendental field.
Quarks and membranes, superstrings
A new theory of everything.

Both weak and strong interaction
Particle and wave diffraction
Building up from many sources
Model fundamental forces
Beauty and simplicity
Reveal quantum gravity
New relationships unfold
From a complex manifold
Quarks and membranes, superstrings
New theories of everything.

Unifying force and matter
Photons and electrons scatter.
Entanglement and quantum state
The paradox of uncertain fate.
Brought together in a Lie Group
From a quark-gluon plasma soup
Unparalled consistency
A beautiful new symmetry.
Quarks and membranes, superstrings
A new theory of everything.

5. Bindings of Ligands in Cells
(God Rest Ye Merry Gentlemen)

Though life appears chaotic and may seem in disarray
Remember that all living things were made with DNA
Except for many viruses, which are made with RNA.

Oh bindings of ligands in cells,
Proteins in cells
Oh bindings of ligands between cells

Though life is so diversified, origins are all the same.
For every living thing on Earth makes use of RNA
For coding and for signaling, and messages to convey.

It's thought that early cells evolved from ancient RNA
And other macromolecules that could self-replicate
And do so more efficiently within lipid membranes

Bacteria are essential in the process of decay
And though it may seem fairly gross,
Do not be led astray
To build lovely new creatures, decay is the only way.

Protists can bind together, and in unison flagellate
And then multiple flagellates, can form an aggregate
Then folding to form tissues as the cells differentiate.

Next time your bread goes mouldy,
Don't just throw it straight away.
Remark upon the miracles that brought you here today.
And witness all the wonder in your decaying baguette.

6. Solar Light (Silent Night)

Solar light, visible light
Sensed by our, power of sight
Part of electromagnetic spectrum
Constant speed when inside a vacuum
Scattered into the colours
Scattered into the colours

UV light, shortwave light
Causes sunburn, when it's bright
Used for sterilization
And forensics irradiation
Flowers reflect it for bees
Flowers reflect it for bees

Longwave light, infrared light
Hot objects, appear bright
Remote controls and night vision
Observe stars and their composition
Snakes sense it to see prey
Snakes sense it to see prey

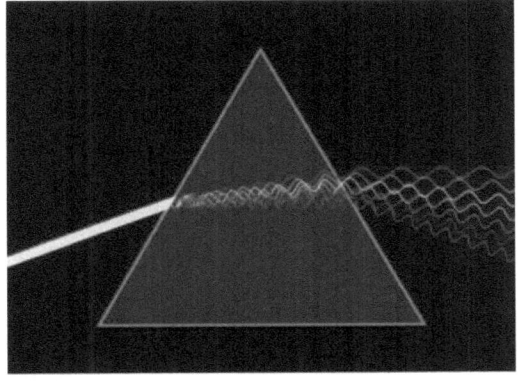

7. Thermodynamics (Blessed Assurance)

Thermodynamics, combined entropy
Statistical mechanics, of heat energy
Functions of work done, measured in joules
And the heat added, moves molecules

Refrain:
Heat always transfers, from hot to cold
Natural disorder, as things unfold
Heat always transfers, from hot to cold
Natural disorder, as things unfold

Perfect equilibrium, homogenous state
Pressure and volume are conjugate
Pressure times volume, plus energy
Is the potential called enthalpy

Refrain

Perfect equilibrium, all is at rest
Energy spread out, all in a mess.
Energy conservation, conservation of mass
The state equations of the ideal gas.

Refrain

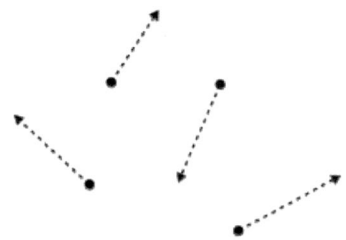

8. Threatened Species (Away In A Manger)

Blue whales are endangered
The baiji is dead
Sadly in that direction
Black rhino's are head

The snow leopard, the panda
And the chimpanzees.
The gorillas and tigers
And the spider monkeys.

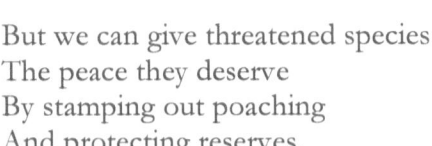

From habitat destruction
Hunting, climate change
From deforestation
For a quick cattle range.

But we can give threatened species
The peace they deserve
By stamping out poaching
And protecting reserves.

It's in our own interests
As we're all interlinked
Not to stand by and watch
As the world goes extinct.

By enforcing fishing quotas
And emissions caps
We can rescue ecosystems
From the brink of collapse.

9. Single Cells (Jingle Bells)

Early heterotrophs
Fed on nucleotides
As their rapid growth
Released carbon dioxide
But autotrophs would rule
As food grew less and less
By generating their own fuel
Through photosynthesis

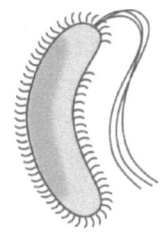

Oh,
Single cells, single cells
Bacteria and algae
Making up about a third
Of world biomass today
Hey
Single cells, single cells
Thrive in every place
Extremes of heat and pressure,
Even in outer space

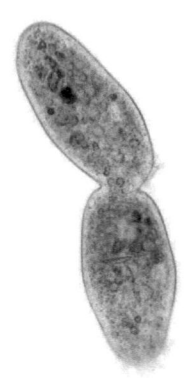

The eukaryotic cells
Are more complex inside
They have more organelles
Especially nuclei
These organelles evolved
As bacterial parasites
Were gobbled up by ancient cells,
And began to thrive inside.

Oh,
Single cells, single cells
Have lipid membranes
Eukaryotes have nuclei
To store their DNA
Hey
Single cells, single cells
Some can flagellate
Amoebas walk with pseudopods
And move by changing shape

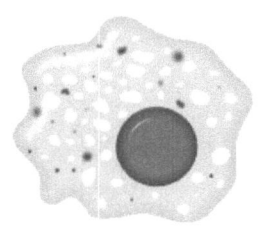

Most species of yeast
Are single celled funghi
On carbs and grain they feast
Making bread and wine
In smaller buds they split
Without a need to mate
But when food sources become scarce
The yeasts will conjugate

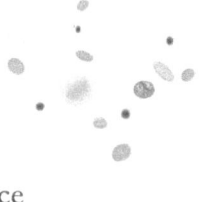

Oh,
Single cells, single cells
Can cooperate
Bacteria form biofilms
Slime molds assimilate
Hey
Single cells, single cells
Most self-replicate
But they have many other ways
To transfer DNA

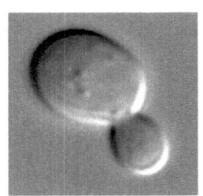

10. Morphogenesis (O Come All Ye Faithful)

One little sperm cell, fuses with an ovum
Forming a zygote with combined DNA
Cleaving and Cleaving. Dividing, not yet growing
Compact by morulation. Hollow by blastulation
Infold in gastrulation - An embryo is formed

Bud little somites
Segment the body axis
A bilateral symmetry most animals share
Grow different tissues, muscle, skin and cartilage
Then fold in neuralation
With neural crest migration
Which helps with heart septation-
Fetus is formed

Optional verse:
Forming the germ layers
Ecto-, meso-, endoderm
Ready to specialize to all types of cell
Grow little organs
Convergent elongation
Oh follow semaphorins
Oh let the water pour in
Through membrane aquaporins-
Baby is formed

Grow little baby
Grow in form and function
Cleaving the spaces between fingers and toes
Follow a program, sort by cell adhesion
So after fertilisation,
And after gastrulation,
And nine months of gestation
Baby is born

11. The Dinosaurs
(Jerusalem "And did those feet in ancient time")

And dino feet, in ancient times
Stomped across Pangea's mountains green
And with majestic Pleiosaurs
Our late Triassic oceans teemed

And yes the long-winged Pterosaur
Glided throughout those humid skies
Until a meteor's, great impact
Led to the dinosaurs' demise

Bringing an age of bitter cold
Bringing a time of untold strife
Did they all die? No! Some evolved.
But there was a great loss of life.

We see them in modern birds
Our world still has the crocodile
Not long ago, the dinosaurs
Ruled over this world for a while.

12. The Big Bang (Little Drummer Boy)

In the big bang everything begun-un-un-un
Infinitely hot and dense in quick expan-sion
While there is still some speculation-on-on
There was exponential infla-a-tion
Infla-a-tion, expansion-on
Slight asymmetry more baryon-ons
Than anti-baryons.

Cooling through a phase transition-on-on
The formation of protons and neutron-on-ons
Then colliding in mass annihilation
The remaining matter was a tiny fraction
Just a fraction, one billion-onth
This all happened in the first second-ond
Not even one.

The first elements formed in nuclear fusion
Hydrogen was built into helium-um
As well as trace amounts of lithium-um
Heavier elements would need stars like the sun-un
Such as carbon, and oxygen
Leaving behind the background radiation
Radiation

Still expanding, gases formed in concentrations
Pulled together by gravitation
As a result of small quantum fluctuations
Towards central black hole event horizon.
Event horizons. Event horizons.
Expanding ever outwards on, and on, on.
Event horizon.
On and on., On and on.

13. Double Helix (We Three Kings)

The four neucleobases in DNA
Form a code for all of our traits
Adenine, Thymine, Guanine, Cytosine
Spiraling all the way

Oh double helix, double strand
With life's information crammed
Strand unzipping, enzyme snipping
Proteins made from short commands

The two strands move in opposite ways
A matching pair for every base
Spiral unwinding, RNA binding
Make messenger RNA

Like a twisted ladder or stairs
Info measured in kilo base pairs
Storing features, of all creatures
From microbes to trees, and bears

The chromosomes are where most genes are
But some evolved from bacteria
Passed down by mothers, and no others
In mitochondria

Codons are bases in groupings of three
Amino acids are coded by these
That's how your genes, make your proteins
From tickertape recipes

14. Fibonacci Sequence (12 Days of Christmas)

On the first day of Christmas, Fibonacci gave to me
A sunflower's arrangement of seeds

On the second day of Christmas, Fibonacci gave to me
A sunflower's arrangement of seeds

On the third day of Christmas, Fibonacci gave to me
2 cauliflowers
And a sunflower's arrangement of seeds

On the fourth day of Christmas, Fibonacci gave to me
3 Artichokes
2 cauliflowers
And a sunflower's arrangement of seeds

On the fifth day of Christmas, Fibonacci gave to me
5 Golden ratios
3 Artichokes
2 cauliflowers
And a sunflower's arrangement of seeds

....and so on until....

On the 12th day of Christmas, Fibonacci gave to me
144 Bees a-buzzing
89 Leaves a leafing
55 Pine cones pining
34 Branches branching
21 Snail shells spiralling
13 Pineapple fruitlets
8 Flower petals
5 Golden ratios
3 Artichokes
2 cauliflowers
And a sunflower's arrangement of seeds

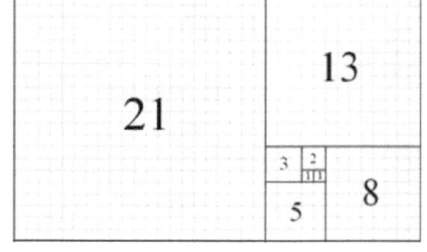

15. Polymers (Ding Dong Merrily On High)

Chains of polysaccharides
Subunits keep repeating
Can be naturally synthesised
Like cellulose and chitin
Pooooooooooooolymers,
Inert and thermally stable
Pooooooooooooolymers,
Insulating and dur-able

New materials we grow
By adding free radicals
Slow to decompose, although
Remarkable materials
Pooooooooooooolymers,
Thermosets and thermoplastics
Pooooooooooooolymers,
Tensile strength and yet elastic

We can change the properties
From chain length, and crosslinking
Long chains have high viscosities
Crosslinks will stop them melting
Pooooooooooooolymers,
A long linked carbon backbone
Pooooooooooooolymers,
Apart from the sili-cones

Chains of polynucleotides
Are helically spiralling
Joined with monosaccharides
Covalently are binding
Pooooooooooooolymers,
Genetic information
Pooooooooooooolymers,
DNA replication

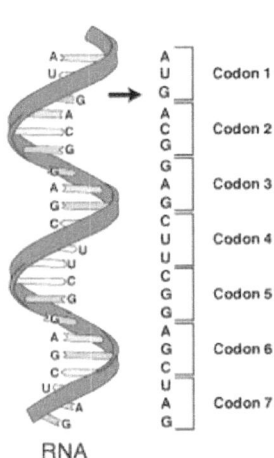

RNA

Ribonucleic acid

16. Neucleosynthesis (Joy To The World)

All living things, All biomass
The fuels, we've used in cars
Were made from fusion, of hot dense gas
Were forged in dying stars
Were forged in dying stars
Were forged, were forged, in dying stars

In the birth of, the universe
Before stars, and galaxies
The clouds of hydrogen, at first dispersed
Were pulled by gravity
Were pulled by gravity
Were pulled, together, by gravity

Stars made all life's ingredients
Apart from hydrogen
The other three, main elements
The carbon and oxygen
The carbon and oxygen
The carbon, and nitrogen, and oxygen

More energy, is needed for
Elements heavier than iron
These are much rarer, they need supernova
Where heavier elements are formed
And heavier elements are formed
And heavier, and heavier, elements are formed

17. Northern Lights
(Angels We Have Heard On High)

Plasma and charged particles
Swiftly streaming from the sun
Intensified in sun cycles
And geomagnetic storm

Nooooooorthern lights, Aurora Borealis
Nooooooorthern lights, Aurora Borealis

Seen mainly in high latitudes
People watching with eyes skinned
Colors at different altitudes
From gases hit by solar wind

Sooooooouthern lights, Aurora Australis
Sooooooouthern lights, Aurora Australis

And the Earth's magnetosphere
Inward leads particle flow
Ionizing the atomosphere
Emits a mainly greenish glow

Nooooooorthern lights, Aurora Borealis
Nooooooorthern lights, Aurora Borealis
Nooooooorthern lights, Aurora Borealis

18. O Chemistry (O Christmas Tree)

O Chemistry, O Chemistry
How numerous are thy branches
Spanning from physics to biology
From materials science to pure theory
O Chemistry, O Chemistry
How numerous are thy branches

O Chemistry, O Chemistry
Thy laws are fundamental
Your reactions portray change
As electrons rearrange
O Chemistry, O Chemistry
Thy laws are fundamental

O Chemistry, O Chemistry
Thy elements intrigue me
With compound structures small and large
And ions with, electric charge
O Chemistry, O Chemistry
Thy elements intrigue me.

19. Macrophages (Rock Of Ages)

Macrophages, protect me
From infection and disease
Guide the bloodstream's monocytes
To my wound, and injury sites
Thoroughly infections cleanse
By engulfing pathogens

For each organ, specalise
And debris phagocytise
From my liver to my spleen
You're a huge, recycling team
Eat the dead cells, to renew
Strengthen bones, and old tissue

Trip the immune system's alarm
To prevent serious harm
Tidy up the neutrophils
Who die after their first meals
First ingest in phagosomes
Then digest, with lysosomes

While I'm fast asleep at night
You continue with the fight
Keeping track of pathogens
By stealing their antigens
Macrophages, protect me
From infection and disease.

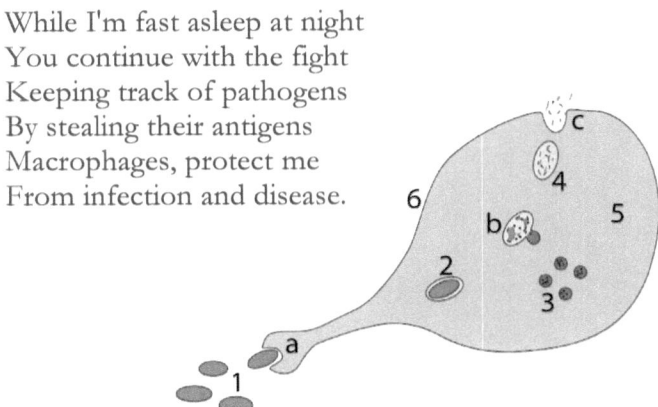

20. Circulation
(Guide Me O Thou Great Redeemer)

Sense pressure, o baroreceptors
And the strict limits maintain
Provide a vital negative feedback
A function of the pressure gain
Circulation, circulation
Deliver nutrients and oxygen
Deliver nutrients and oxygen

Constrict now all my blood vessels
Making blood harder to flow
By increasing vascular resistance
Blood pressure won't get too low
Strong deliverer, strong deliverer;
Take CO2 and wastes away
Take CO2 and wastes away

Every cell needs oxygenation
To turn nutrients into energy
Essential are the mitochondria
They produce the ATP
Respiration, respiration
Cycling ATP to ADP
Cycling ATP to ADP

21. Energy (Nearer My God To Thee)

Oh what is energy, what's energy?
It's an ability, a scalar quantity
In things that move freely
Or stored potentially
Everything is energy, we are energy.

And when a mass is held, up somewhere tall
Stored in its tendency, suddenly to fall
There gravitationally
Is potential energy
All mass is energy, mass is energy.

Whether you measure in, joules or calories
Food contains energy, chemical energy
And when the bonds release
Their potential energy
The cells in you and me, use the energy.

And we can think of heat, as thermal energy
Molecules move more quickly, with a heat increase
So thermal energy
Is kinetic energy
All heat is energy, kinetic energy

And if we look closely, protons and neutrons
Are just three quarks which means, they are baryons
And their mass energy
Is potential energy
All mass is energy, potential energy

From relativity, it equals mc squared.
And at the greatest speeds, energy is shared
When going very fast
Things increase in their mass
Kinetic energy. potential energy

22. Dear Large Collider
(Dear Lord and Father of Mankind)

Dear large collider based at CERN
Crash protons at great speed
That we may new quanta discern
And build the theories when we learn
From the observations we need
With confidence succeed

The laws of particle physics are
Quite tricky to depict
The standard model worked so far
And though the rules can seem so bizarre
New particles it predicts
Without any conflicts.

A canister of hydrogen gas
Stripped of its electrons
Accelerated oh so fast
The protons increase in their mass
And in their collisions
Producing new bosons

The great consortium of mankind
Working together in peace
And furnished with the finest minds
Who knows what new physics they will find
The final jigsaw piece
Of supersymmetry.

Search for the truth of gravity
Its particle and field
A model that consistently
Predicts all of the facts that we see
And the mysteries revealed
In the data flood we yield.

23. Global Warming (Carol of The Bells)

Global Warming,
It's now coming,
All seem to say,
"It's here today",
Long droughts are here
For us to fear,
Flooding as well,
As the seas swell

Tick, tock, tick, tock
That is the clock
Time's running out
Without a doubt
Gas trapping heat
Melting the peat
Releasing more gas
Warming up fast

Oh the feedback
The time to act
Was yesterday
But don't dismay
No time to waste
Act with great haste
Green energy
Technology

Global, global, global, global warming
Global, global, global, global warming

Time to invest
Save rainforest
This is our chance
Time to advance

24. How Peculiar (Hallelujah)

I've heard that when Charles Goodyear
A 19th century Engineer
Was trying to make rubber durable, in the winter.
And well, so the story goes
He dropped it on
An open stove
It toughened, and he thought
"well how peculiar..."

How peculiar, how peculiar, how peculiar,
how pecuuuuuuliar....

Switch to 1938, Plunkett was trying to replace
The gasses in a home refrigerator
But after the new gas had gone
It left behind some pure teflon
The perfect non-stick surface, residue yeah.

How peculiar, how peculiar, how peculiar,
how pecuuuuuuliar....

The year was 1968, 3M were trying to create
An extra-strong adhesive super glue yeah
They failed, but goes the anecdote
They created, the post-it-note
By applying this failed glue to sheets of paper

How peculiar, how peculiar, how peculiar,
how pecuuuuuuliar....

Well when the church insists the Earth
Is centre of the Universe
Well you don't really argue with them, do you?
But what if you should know truth
Cos you've watched the stars, and seen them move
You've even seen 4 moons go round Jupiter.
How peculiar, how peculiar, how peculiar,
how pecuuuuuuliar....

When Flemming took some time away
And purely accidentally,
Left a few Petri dishes on the counter.
On his return, he found in both
An orange Penicillium growth,
And the fungus had killed all his bacteria.

How peculiar, how peculiar, how peculiar,
how pecuuuuuuliar....

Image: Moons of Jupiter: Jan Sandberg, www.desert-astro.com

25. Radioactivity (Good King Wenceslas)

Henri Becquerel did glance
On his plates sincerely
Left in dark draws pure by chance
Yet they'd been marked clearly
Uranium crystals emit rays
Emit rays quite freely
This finding would be appraised
By young Marie Curie

Marie:
"Come and watch this Pierre my dear
Your electrometer
Uranium's signal's marked out clear
The ore is yet still greater
That suggests new elements
Truly most unstable
We'll refine this raw pitchblende
Expand Mendeleev's table"

"Bring me ore and bring me gas,
Bring me iron rods hither
Thou and I shall form a mass
Crystalised in slivers"

Intensely working as a team
Soon they'd be preeminent
Through the acid's noxious steam
Found they two new elements

"How the world grows darker now
With disquieting silence
War is looming, but somehow
I will use this science"

Driving in an x-ray van
Through the dreadful slaughter
Helping every shattered man
Marie and her daughter
In her life's determined will
Marie sought out answers
Her Radium, we use it still
For treating certain cancers
Therefore you may rest assured
Whatever you are dreaming
Life will be its own reward
Search for your own meaning.

26. Tectonic Plates (Amazing Grace)

Tectonic Plates, they move around
And change the boundaries
Where once were oceans, Now there's ground
And mountains, Now are sea

'Twas plates that made the atlantic ridge
And plates the land revolved
Creating and destroying bridges
Plates changed how life evolved

Though where plates meet with great friction
Can cause unstable zones
Life may have never first begun
From solar heat alone

Though moving boundaries cause earthquakes
And sometimes volcanoes
Our world was shaped from moving plates
Millions of years ago

27. Special Senses (In Dulci Jubilo)

The pressure waves in air
Reach ear canals, and there
Detected through vibration
By three most tiny bones
Transmit this information
About all of these tones
Amplitude, frequency
Into nerve activity

And in the inner ear
Not only used to hear
Balance and orientation
And in the vestibule
Angular acceleration
The fluid-filled saccule
A bony labyrinth
Filled with endolymph

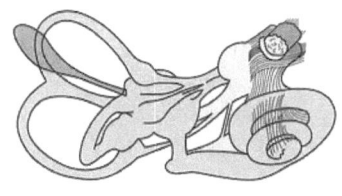

And the sense of sight
Perceives reflected light
Sensed by photoreceptors
The rod cells and the cones
The cones give us the colors
The rods work in the dimness
Synapse biploar cells
Make graded potentials

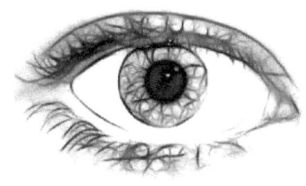

We sense both tastes and smells
From chemosensor cells
When things taste sour or salty
that's ions moving through channels
but bitter, sweet and umami
Are ligand binding particles
A symphony of smells
Hundreds of specialised cells

28. Electrocardiogram
(O Little Town Of Bethlehem)

Oh the Electrocardiogram, is used to analyse
The waves and wells, Of pacemaker cells
As they depolarise
And from tiny electric charges
Events as they arise.
The posterior test, for cardiac arrest
That's helped so many lives.

And used as well in toxicology
To find arrhythmia
The vital checks, on drug effects
The baseline and QR
Drug binding to receptors,
On membrane ion channels
Of potassium, and sodium
That change the charge of cells

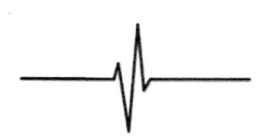

When metal ions, flow in and out
They make a wave of charge
The natural feat, in each heart beat
Of creatures small and large.
And with so many applications
From research to critical care
And home ECG, through telemetry
That's small enough to wear

Your medulla increases your heart rate
When you do exercise
By releasing, adrenaline
To make your blood flow rise
Baroceptors measure pressure changes,
Send your brainstem this news
And all to feed, the oxygen need
Of organs and tissues.

29. Solar Cells (The First Noel)

The first solar cell, in 1839
Was a conductive solution of silver chloride
This solution, was where, we were first to detect
Light creating a current, the pholtovoltaic effect
Solar cell, solar cells, solar cells, solar cells
First discovered by Edmond Becquerel

When an energized photon, Hits a free electron
It may be released as a photoelectron
And with few electrons, in their outer shells
Semiconductors and metals, release them quite well
Outer shells, outer shells, outer shells, outer shells
Causing the power of solar cells

There are many ways to increase, PV efficiency
Like doping the two layers, with impurities
And adding phosphorus, gives a spare electron
While a hole is created by adding boron
Boron, boron, boron, boron
Creating the p type layer of silicon

And the current flowing out, from n back to p
Is a flow of electrons, called electricity
And conducting in the sun, like a photodiode
But able to do work, and power a load
Solar cells, solar cells, solar cells, solar cells
One of the central renewables

30. What Goes Around
(Down In The River To Pray)

As I sat down in a pretty sunset
Reflecting how it's easy to forget
The world is really spinning round
That's why the sun appears to set

Oh the world it spins around
Spinning around, and around
Once a day it spins around.
Causing the night and the day.

As I looked up at the moon in the sky
Remembering the real reason why
It changes shape, from thin to round
In phase with the changing tide

Oh the moon it goes around
Once a month, it goes around
The side we see as it goes round
Is partly lit by the sun's rays

As I basked in the hot summer's sun
Remembering how this warm season
Is because the earth, is going around
The north's near, then far away

Oh the world it goes around
Round the sun, it goes around
Once a year it goes around.
Causing the seasons to change.

As I looked up at the glistening stars
Thinking that they are just oh so far
But gravity makes, us all go round
All around the same way
Oh the galaxy goes around
To a black hole we all go round
200 million years around
Fast, but it's such a long way

As I went down in the underground caves
Studying about the cosmic rays
And looking for the neutrinos
From the sun and supernovae

Oh neutrinos streaming down
From the sun, come on down
Tiny particles come on down
Bombarding us every day

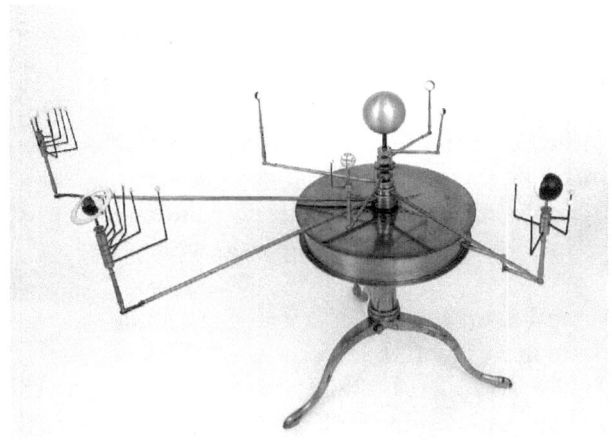

31. We Are Part (How Great Thou Art)

When I at night, look up in awesome wonder
And ponder on the vastness of it all
I feel the rain, I hear the waves and thunder
And feel a beauty that continues to enthral.

Refrain:
And feel involved with everything I see
For I am part, and we are part
I'm in the world, and it's inside of me
For I am part, it thrills my heart.

I am amazed, nearly all the atoms in my body
Will be replaced, within a single year
In constant change, which all our lives embody
I feel so free, from all my guilt and fear.

And in my life, when I new friends encounter
I give them all my love, and deep respect
They too are part, of all the awesome grandeur
And in chemistry, our separate lives connect

And when I think, the world will keep on turning
And how the universe won't notice when I cease
Time will move on, the stars continue burning
And suddenly, I'm overwhelmed with peace

When I want to share the way I'm feeling
There is so much, I scarce know where to start
The joy and love, with all my senses reeling
And then proclaim, "the world, how we are part!"

32. The Scientific Method
(I Vow To Thee My Country)

The scientific method is
An ordered search for truth
A way of answering questions
Through evidence and proof

The proof that is sufficient
The proof that stands the test
That discounts rival hyphteses
And leaves only the best

The proof that can be falsified
In tests we can repeat
The proof not contradicted
Which makes theory complete

And though there may be many things
That we will never know
It is only through this process that
Our reasoning will grow

Each discovery brings more questions
It can get more confusing
But we develop new technology
Through each advance we bring

And proof, by proof, and tirelessly
Our knowledge does increase
And our ways, are ways of curiousness
And all our paths are peace.

33. Blue Marble
(Land of Hope and Glory)

Planet of cloud and ocean,
Mother of the green
Locked in cyclical motion,
of tilt and eccentricity
Protecting us from radiation,
precious thin atmosphere
Leading many to question.
Why, oh why, are we here?
Leading many to question.
Why, oh why, are we here?

Spinning on your own axis,
causing night and the day
In the periodic blackness,
we see the milky way
Of all other known planets,
the only with living beings
Is the one we inhabit, Are we so unlikely?
Is the one we inhabit, Are we so unlikely?

Breathtaking are your vistas,
boundless seem your skies
Spectacular pink sunsets,
met by orange sunrise
We call you the blue marble,
Cradle of the green
Maybe it's sentimental, But you are beautiful to me
Maybe it's just sentimental. But you, are beautiful,
to me.

34. First they swell, then explode
(It is well with my soul)

When stars are burn'd out, at the end of their days,
Expanding they less brightly glow,
High mass stars blow up, as bright supernovae
First they swell, First they swell, then explode

Refrain:
First they swell, then explode
First they swell, first they swell, then explode

And blasting material every which way
A shockwave that ripples and rolls
The elements they make, are in us all today
The biggest stars', cores collapse, to black holes

Refrain

At the centre of most every galaxy
Lies a supermassive blackhole
The singularity, with infinite density
Pulls us all, t'wards itself, we revolve

Refrain

35. O Photosynthesis
(Great is Thy Faithfulness)

O photosynthesis, maker of glucose
Taking up water and light energy
Changing the CO2, to carbohydrates
Freeing the oxygen that we all breathe

Refrain:
O photosytnthesis, O photosynthesis
Supplier of nearly all life's energy
Working, reacting, in light and in darkness
Making our food, and the air that we breathe

Algae and seaweed, bacteria to mighty trees
Trees absorb mainly through specialised leaves
Fixing the carbon inside their own bodies
Building themselves, with a rich chemistry

Refrain

Messy processes evolved not created
So many steps that are not optimised
Plants would be black if they used the whole spectrum
Still it's the basis of nearly all life

Refrain

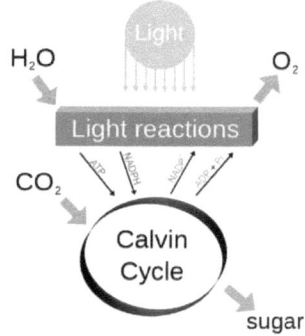